# DON'T BELIEVE IN GOD?

# HERE'S THE BOOK THAT WILL CHANGE YOUR MIND

**Lawrence Newman**

Don't Believe In God?

Copyright © 2021 by Lawrence Newman

ISBN 978-1-7347100-3-8

Library of Congress Control Number: 2021930537

Printed in the United States of America

First Edition

Publisher: Silver Millennium Publications, Inc.
               Gold Canyon, Arizona

All rights reserved. No part of this book may be reproduced in any form, or by any means electronic, mechanical, recording, or otherwise, without written permission from the author, except for brief quotations used in reviews.

*The Creation of Man--Michelangelo*

**Dedicated to all those looking
for the proof of God's existence**

# TABLE OF CONTENTS

**Introduction**

**Chapter 1 : Why Something Instead Of Nothing?**

**Chapter 2 : The Big Bang**

**Chapter 3 : Intelligent Design and the Laws Of The Universe**

**Chapter 4 : The Miracle of Life**

**Chapter 5: The Ultimate Creation**

**Chapter 6 : Some Phenomena To Consider**

**Chapter 7 : Alone In The Universe?**

**Chapter 8 : The Devil's Advocate**

**Chapter 9 : What Kind Of God?**

**Chapter 10: Putting It All Together**

**Chapter 11: Final Thoughts**

**Relevant Quotes**

**Reference Sources**

# INTRODUCTION

I was an agnostic for most of my adult life. About ten years ago I began studying material arguing for and against God's existence. Many atheists present strong arguments for their position, including the late Christopher Hitchens, who I consider a tremendously gifted writer who died far too young, and Richard Dawkins, a polished speaker and writer, who is the leading proponent of atheism. Most of their arguments have to do with the pain and suffering they see in the world caused by man or natural causes that would not exist if there was a merciful God.

Man has debated God's existence for thousands of years. Many looked around and concluded that what they saw had always existed in some form. As scientific inquiry advanced the wonder of the discoveries concerning the very small i. e., the atom, and the very large i. e., the universe, seemed to point to another possibility—an intelligent designer at work. Many others have traveled the same road I have, searching for answers about God's existence. I have incorporated many of their thoughts, through quotes, in this book.

The evidence I gathered in my search for the proof of God's existence is presented in the following chapters. Although not any one piece of evidence presented may be persuasive to everyone, it is the accumulation of the evidence that led me, and may lead the reader to come to the same conclusion. Much of the material, especially that dealing with astronomical and other scientific information, may be known to many of the readers, but it was necessary to present it so that all readers would be made aware of it.

It is not possible to prove God's existence or non-existence beyond all doubt. Only if one personally heard God's voice from heaven or experienced a miracle in the form of a "burning bush" or "parting of the Red Sea" moment, would God's existence be proven to that person. On the other hand it is also difficult to prove the nonexistence of an entity that is invisible to human eyes.

Throughout the text I refer to "man". In the context of this subject, "man" is an all-inclusive noun without any gender significance—it is a human being possessing self-awareness coupled with intelligence that sets him or her apart from all other living things.

Everyone must come to their own decision regarding God's existence. The journey I experienced is laid out in the following pages. I hope this book, describing the information I reviewed during my journey, is helpful to others who are thoughtfully examining their own beliefs.

# CHAPTER 1

## WHY SOMETHING INSTEAD OF NOTHING?

The first issue to be discussed concerns an examination of the question "Why is there something instead of nothing?" Some might consider the "why" of this question more applicable to a philosophical discussion, however this question will be initially examined from a "cause and effect" standpoint. The philosophical aspects will be discussed later.

The concept of nothingness can be difficult to define, although if asked, most people would probably describe a condition devoid of all sound and light, i. e., silent blackness.

Since there obviously is "something" then one would question how that "something" came into existence. There must have been a cause which created the "something". It defies all logic that material things would spontaneously come into existence without a cause.

Once one accepts that there was a cause that brought the universe into existence, then one can begin to examine what that cause could be—and the physical laws that govern the existence of what we see today.

*"Science cannot answer the deepest questions. As soon as you ask why there is something instead of*

*nothing, you have gone beyond science. I find it quite improbable that such order (the observable universe and the physical laws underlying its existence) came out of chaos. There has to be some organizing principle. God to me is the explanation for the miracle of existence—why there is something instead of nothing."—Cosmologist Allan R. Savage*

A person skeptical of this conclusion would then ask, "If God created the universe, then who created God?" One would seem to be facing an unanswerable query. However, after deeper consideration it would seem obvious to most of those pondering this question that God, by definition, does not need a creator. He, or it, is the first step in all causal events that follow—the First or "Prime" Cause.

Can the human mind conceive of such an entity—something that has existed for eternity and is capable of bringing into existence a universe of unparalleled complexity? Only with great difficulty, if at all. I do not pretend that I understand the mind of God but I can present the evidence that I believe supports God's existence.

Science has determined that there are physical laws governing the "something" that surrounds us. If these laws or the relationship between them were changed, even to the slightest degree, the universe, and life here on Earth would not be possible.

When and how did all this begin?

## CHAPTER 2

## THE BIG BANG

At one time it was believed by most scientists that the universe had always existed. Even Albert Einstein, the 20th century's leading physicist, initially believed this. In his theory of general relativity attempting to explain the physics of the universe he had inserted a "cosmological constant" to ensure the universe was stationary. It was only in 1931, after he met with Edwin Hubble, an American astronomer, who proved to him, by astronomical observation, that the universe was expanding, that Einstein proclaimed his formulation of the "cosmological constant" the "biggest blunder of my career". The concept of an expanding universe was difficult for him and many atheistic scientists to accept as it implied a beginning—a moment of creation.

By measuring the universe's expansion and working backwards, scientists were able to determine that the universe came into being approximately 14 billion years ago in a "Big Bang"— an explosion of light and heat of 100,000,000,000 degrees Centigrade. In essence, the universe and all the laws of its existence, including those that would lead to life here on Earth , were created from nothing by some supernatural force.

The existence of everything we see today, including the multitude of galaxies visible through telescopes, was set into motion in that cosmological instant.

The calculated strength of the explosive force of the Big Bang turns out to be miraculous. If the force of explosion had been just a little stronger the material of the universe would have been blown into an expanding, diffused soup of particles that could not have formed the galaxies, stars and planets. If the force of the explosion had been slightly weaker, the material would have been pulled back by gravitational forces in a very short time, on a cosmic scale, and imploded. It is calculated that the force of the explosion measured by the ratio of matter and energy to the volume of space at the instant of the Big Bang had to be within one quadrillionth of 1 percent of ideal—the mathematical equivalent of having a thrown penny come up heads billions of times in a row.

Since the Big Bang occurred the universe has expanded, forming billions of galaxies, each containing hundreds of billions of stars. One scientist, astrophysicist Robert Ciconne, stated, "The estimated number of stars in the universe is equal to all the grains of sand on all the beaches on Earth." In other words, it's a number unable to be grasped by the human mind. And remember, this all started from a point in the silent blackness where nothing existed.

The size of the present known universe can be roughly estimated by using the concept of light years—a light year being the distance light will travel in one year—at a speed of approximately 186,000 miles per second—or 5,878,625,000,000 miles—give or take a couple of hundred thousand miles. Using this basis of measurement it is estimated by astronomers that the size of the observable universe, end to end, is 93 billion light years.

The Milky Way galaxy in which our solar system is located contains 100-400 billion stars and is approximately 110,000 light years across. The largest galaxy presently known in the observable universe contains approximately 100 trillion stars.

The distance from our Milky Way galaxy to the Andromeda galaxy, the nearest galaxy to ours, is approximately 2 ½ million light years. Other galaxies, numbering in the billions, stretch out as far as our present astronomical equipment can see. Some astronomers think the universe is infinite since it is expanding at an accelerating rate.

Over time many stars have been born and died within our galaxy. Two chemical elements comprise the major components of a normal star—hydrogen and helium with some trace amounts of carbon, iron, oxygen, neon and other elements. However, when a star ages and dies it produces substantial amounts of an element that is essential for life—carbon, and other elements. These dying star elements are spread throughout the galaxy and incorporated into planets that surround other stars.

In essence, what we see here on Earth, including our own bodies, is composed of the remnants of stars that died long ago.

The nearest star to the sun, is Proxima Centuri, located about 4 ½ light years from us. If we considered the size of our sun to be approximately the size of a grapefruit located in New York, Proxima Centuri would be located in Los Angeles—about 2500 mile away. The Earth's size, comparatively speaking, would only be a dot on the grapefruit.

The astronomical information in the preceding paragraphs has been presented in order to allow the reader to consider the magnitude of the known universe and the distances involved since its creation 14 billion years ago. The concept of space travel to other stars, in search of planets with intelligent life, given the immense distances involved, seems very improbable. Putting aside Hollywood's ability to propel men to other parts of our galaxy by "warp speed" or through "wormholes", man's ability to move a spacecraft even a small percentage of the speed of light, considered by scientists to be the outermost theoretical speed limit, is to be found nowhere on science's drawing boards. The space satellite, Voyager I, which was launched in 1977, currently traveling at 36,000 miles per hour, has just recently left our solar system. If it continues at its present speed it will reach Alpha Centuri in 80,000 years. Andreas Hein, an engineer with the Icarus Interstellar Project, developed an estimate for an interstellar trip. His conclusion: Such a mission would cost over $174 trillion and take 40 years of development. The mission would be an unmanned 50-year journey to Barnard's Star (about 6 light years distant) using a spaceship with "fusion drive" technology to approach 12% of the speed of light, consuming 50,000 tons of fuel over that period.

Even allowing for the fact that "fusion drive" may never get off the drawing board, it's doubtful society would want to spend an amount equal to approximately seven times the current U. S. national debt, considering all the problems we face here on Earth The cost and time involved makes the concept of interstellar travel look like a pipedream. Our future is here within our own planetary system—and specifically here on Earth. Let's hope man does not destroy this oasis of life, since there is nothing even remotely like it anywhere that we are aware.

*"If you turn on your television and tune it between stations, about 10 per cent of that black-and-white speckled static you see is caused by photons left over from the birth of the universe. What greater proof of the reality of the Big Bang—you can watch it on TV."*—Jim Holt from his book "Why Does The World Exist: An Existential Detective Story".

## CHAPTER 3

## INTELLIGENT DESIGN AND THE LAWS OF THE UNIVERSE

The more scientists attempt to unravel the mysteries of the universe—from the largest (on a cosmic scale) to the smallest (components of the atom), the more they are amazed by what they find. The vastness of the universe was discussed in the preceding chapter. A discussion of the physics of atoms and the physical laws that underlie the existence of matter follows.

Look around your present environment and consider this: Everything you see is composed of atoms that are incredibly small. It is estimated that in one grain of sand consisting of molecules[2] of silicon dioxide ($SiO_2$) there are 78 quintillion (78,000,000,000,000,000,000) atoms, 2/3 of which are oxygen atoms and 1/3 of which are silicon atoms.

The composition of each of these atoms is a nucleus of protons and neutrons surrounded by electrons circling the nucleus at incredible speed. The electromagnetic force holding each atom together is amazingly powerful and scientists only recently (within the past 100 years) are beginning to understand the underlying physics of atoms. Huge, very expensive, laboratories have been built throughout the world, which are attempting to discover the inner working and components of the atom. These laboratories, housing particle accelerators, have uncovered many sub-particles of atoms. Science is finding that unlocking the secrets of the atom are like taking an onion apart, layer by layer, in an attempt to identify the incredible complexity of what exists within Atoms have amazing properties. One of these properties is their ability to vibrate at incredible speeds. An atom of the element cesium vibrates at over 9 trillion times per second, allowing us to calibrate clocks to within one second every 30 million years!

*"If, in some cataclysm, all of scientific knowledge were to be destroyed, and only one sentence passed on to the next generation of creatures, what statement would contain the most information in the fewest words? I believe it is the atomic hypothesis (or the atomic fact, or whatever you wish to call it) that all things are made of atoms—little particles that move around in perpetual motion, attracting each other when they are a little distance apart, but repelling upon being squeezed into one other."—Physicist Richard Feynman*

Although many people are familiar with Einstein's famous equation $E=mc^2$ the ramifications of that simple equation are only known by a small minority of the public. What the equation means is that energy and matter are interchangeable. The few ounces of matter that were annihilated and turned into energy over Hiroshima on August 6, 1945 only begins to scratch the surface as to the total energy contained in all the matter in the universe. It is calculated that the nuclear energy in a single gram of matter, less than the weight of a penny, could lift a million-ton load six miles high. The most astounding fact about that equation is that it works in both directions—not only does the annihilation of matter result in the release of a fantastic amount of energy it also means that solid matter can be created by the input of that same amount of energy. Pure energy, in an almost incomprehensible amount, created the basis for all the matter now in existence at the moment of the Big Bang.

Consider the following physical laws of the universe:

 . . . the ratio of the mass of the proton to the electron in an atom is 1,836,153 to 1—the exact ratio needed for matter to exist.

 . . . the mathematical value of the gravitational force ($6.67384 \times 10^{-11}$) is set exactly right. It could crush us if it were stronger. If it were weaker the universe would not hold together.

 . . . the mathematical value of the electromagnetic force (1/137.035999) that underlies all chemical bonds is set exactly right. If it were different it would cause us to become non-existent.

... the mathematical value of the strong nuclear force is set exactly right. This force is responsible for binding together the fundamental particles of matter to form larger particles. If it were different atoms would collapse or explode.

... the mathematical value of the weak nuclear force is set exactly right. If it were different many atoms would become radioactive and stars would not shine.

*"Why nature is mathematical is a mystery. The fact that there are rules at all is a kind of miracle."—Richard Feynman, a Nobel Prize winner for quantum electrodynamics*

Scientists now agree that even very small alterations in these values would have resulted in catastrophic changes in the resulting universe. If even the relationship between these values had been slightly different the universe as we know it would have been nonexistent and life impossible.

The Anthropic (Man) Principle was first put forward in 1974 by cosmologist Brandon Carter in a lecture to the International Astronomic Union. He drew attention to a number of astonishing "coincidences" among the physical laws of the universe. He concluded that infinitesimal changes in the mathematical values of any of these laws would have resulted in a universe profoundly different from our own—and devoid of life.

The conclusion that Carter and many scientists now have come to is that the values of the physical laws of the universe are set exactly right to support an environment that is conducive to life as we know it here on Earth.

# CHAPTER 4

## THE MIRACLE OF LIFE

The human genome (DNA) contains approximately 3 billion bits of information, consisting of the nucleotides adenine, cytosine, guanine and thymine, expressed by the letters A, C, G and T. It is a blue print for the construction and functioning of the human body. In other words, someone or something set down an intricate plan that controls every function of the human body. The plan existed before the functioning of the body, that is human life, could begin. Who created this information—a huge set of commands?

*"DNA is like a computer program but far, far more advanced than any software ever created."*—Bill Gates, co-founder of Microsoft

Many scientists now believe that mere chance or evolution alone could not have been the basis for the development of this DNA blueprint and the construction of a human body, or for any living thing, whether plant or animal.

*"The chance that higher forms (of life) have emerged in this way (by evolution) is comparable with the chance that a tornado sweeping through a junk yard might assemble a Boeing 747 from the materials therein."*—Fred Hoyle, Professor of Astronomy at Cambridge University

Further, Hoyle also compared the likelihood of obtaining a single functioning protein by chance to a solar system full of blind men solving Rubik's Cubes simultaneously.

From the smallest one-celled living thing to the multicellular human being, all are masterpieces of engineering. The probability of life just coming into being, without a guiding hand, is impossible to envision.

When studying the intricate engineering of living things the concept of irreducible complexity comes into play. Irreducible complexity can be most easily understood when discussing machines—for example, a car. If you take away any of its major components—the tires, the chassis, the engine, the electrical system, the fuel system, etc., you would not have an operable car. In other words the entire car has to come into being by fabricating it at one time. It could not evolve in stages. Such is the irreducible complexity of living things—they had to come into being as a whole, not in parts. One-celled animals do not evolve into a dinosaur or a human being. Intervention, in the form of a blueprint maker, is required.

All scientists who have attempted to create life from lifeless chemicals have failed. There is no foreseeable process that can construct a human being with its astonishing complexity. Living things are separated from non-living things by an unbridgeable chasm.

Life, in its most primitive form, arose hundreds of millions of years ago, in the Pre-Cambrian Era. Evidence of soft membrane worms and simple cellular life can be found in the geographical strata of that time. Suddenly, approximately 540 million years ago there was an explosion of life in the oceans—the Cambrian Explosion. A multitude of various forms of life appeared in the ancient seas of that time. The important point to be made here is that this "explosion" of life was just that—almost immediate—at least in terms of the time frame of Earth's development. There was no gradual evolution of life forms over eons. Charles Darwin, and the scientists who followed, were, and are, at a loss to explain this explosion within the context of evolution.

Over the passage of time animal and plant life migrated to land from the oceans and reptiles rose to prominence. The dinosaurs were absolute kings of the planet for hundreds of millions of years. Small mammals, in the form of rodents, which existed at the time, seemed to be unlikely challengers to the reptiles that ruled the land. And then, an astronomical event occurred which changed life on Earth forever.

Approximately 65 million years ago an extinction event occurred that cleared the planet of almost all of its reptilian life—a large meteor struck the Earth with tremendous force at the edge of Mexico's Yucatan Peninsula in the Gulf of Mexico. The explosion, estimated to have packed the power of two million hydrogen bombs, caused a 110 mile wide crater. The effects of the meteor's impact killed off the dinosaurs and two thirds of the other animal and plant species on the planet. It is thought that a plume of material from the explosion and the smoke from the subsequent widespread fires circled the Earth causing a die-off of much of the plant life, the food source for many of the planet's inhabitants, and that temperatures were lowered considerably for many years. The important point to be made here was that the dire effects of the meteor's impact were set at just the right level to kill off the dinosaurs but not all of the warm-blooded mammals or plant life—similar to the miraculous force of the Big Bang. There was just enough life remaining to begin anew, and climax in the rise of mammals—including man.

The plant and animal life that died off over a period of millions of years, before and after the extinction event, became the coal, gas and oil that provided our age with the energy to warm our homes, cook our food, power our factories, and power the aircraft, ships and vehicles we use today. This abundance of potential energy produced underground by the decay of the life eons ago was fortuitous. Consider how human life on the Earth would have been different if these sources of energy had not been available.

# CHAPTER 5

## THE ULTIMATE CREATION

We will now consider the complexity of man—the senses and some selected organs of the human body—hearing, sight, smell, touch, speech, the brain and the cardiovascular system. As mentioned previously, these systems are miracles of biological engineering—with a DNA blueprint of 3 billion pieces of code, directing how the human body is to be constructed and how it is to operate—in the most minute detail.

### HEARING

By the transmission of sound waves against the human ear drum and the movement of that sound along an auditory nerve to the brain, man is able to relate to the outside world in a way that even sight can't match. It has been judged that between man's sensory organs that transmit sight and sound, it is sound that gives him more feeling of being a part of his environment.

In addition to being an essential part of our everyday communication with others, we have the ultimate pleasure of experiencing the beauty of music. Whether listening to the melodious compositions of Bach, the symphonic masterpieces of Mozart and Beethoven or the pulsating rhythmic score of a Hans Zimmer film soundtrack, the experience of hearing the beauty of music can be a soul-piercing event.

Some consider music a gift from God:

> "If I should ever die . . . let this be my epitaph: THE ONLY PROOF HE NEEDED FOR THE EXISTENCE OF GOD WAS MUSIC"—Kurt Vonnegut

> "Music, the greatest good that mortals know. And all of heaven we have below."—Joseph Addison

> "Without music, life is a journey through a desert."—Pat Conboy

> "Whether I was in my body or out of my body as I wrote it (the Messiah's 'Hallelujah Chorus') I know not. God knows."—George Frederick Handel

Music truly is a gift from God. Consider this: In order to experience this gift God had to instill in humans by way of our DNA three abilities—first, the ability in a select group of humans to compose beautiful music; secondly, the ability of another group to master the playing of musical instruments to play the composed music; and finally the ability of humans to appreciate the music when they hear it. Remove any of the parts of this divinely inspired musical triangle and one of the greatest joys of living is lost.

SIGHT
The human eye is another miracle of biological engineering, taking in images, regulating their intensity, focusing them through lenses and then converting the

images into electrical signals before transmitting them through the optic nerve to the brain. It is estimated that 10 billion calculations take place every second in the retina in the process of transmitting the light image to the brain. In addition to differentiating between approximately seven million colors the human eye can handle a range of light 10 billion to one; the best made photographic film can only handle a range of light of 1,000 to one. As with the ear, the eye's functions are laid out in a DNA blueprint put in place before the eye came into existence.

Sit back in your chair and consider all the information that is cascading into your eyes—millions of impressions of color, form, distance, meaning and relationship—all processed instantly through a time spectrum of your past experience with those impressions. In addition to giving you the ability to function within your environment eyesight gives you the ability to appreciate the beauty of the world around you—whether it is fields of wildflowers, rainbows, snow-capped mountains, the Grand Canyon, a beautiful painting or the grandeur of sunsets and sunrises.

*"The next time a sunrise steals your breath away or a meadow of flowers leaves you speechless, remain that way. Say nothing, and listen as heaven whispers, 'Do you like it? I did it just for you."*—Max Lucado

## SMELL

Our sense of smell is another sensory miracle that man possesses. Although the human sense of smell is not nearly as strong as that of many animals it is still capable of providing man with the ability to differentiate between thousands of different smells.

The function of smell is performed by two odor-detecting groups of cells numbering between five and six million located in the nasal passage. Other animals contain substantially more of these cells, such as a dog, whose cells number over 200 million. It has been said that although you may smell a cake baking a dog can smell the butter, the sugar, the chocolate, flour, the eggs and other ingredients baking separately. Owls, on the other hand, have a very poorly developed sense of smell, and are one of the few predators that hunt skunks.

Our nose is also one of the main organs that control taste. Although the tongue can recognize the qualities of sourness, sweetness, acidity, saltiness and savorieness, it is the nose that overlays these qualities with smell that give taste to our food.

Some smells warn us of potential harm like smoke and harmful gases. But think how bleak our lives would be if we could not smell the flowers around us, the countryside after a rain, food and perfumes. One of life's pleasures for many is walking through the toiletries section of a large department store, inhaling the delightful blend of aromatic scents.

## TOUCH

While your other four senses are located in specific parts of the body, your sense of touch is all over your body since it originates in your skin. The dermis portion of your skin is filled with nerve endings that transmit information to your brain about the things which your skin encounters. When the touch, pain or heat sensors in your skin are stimulated, they send electrical pulses to special cells that relay electrochemical impulses to your spinal cord, which in turn sends it to your brain. Pain receptors are probably the most important to you because they warn you when your body is being hurt. The transmission of pain takes place very quickly. Once your hand picks up the hot handle of a pan you immediately let go.

The sense of touch also conveys pleasure in many ways, such as the flow of water around your body while taking a shower or swimming, the feeling of breezes against your skin or the warmth of a crackling fire. However, it is in the interaction with other living things that the sense of touch is most pleasurable, whether it is the cold nose of a dog or the warmth of the body of another human being—and reaches its highest sensory expression during the sexual experience.

## SPEECH

The human vocal cords are composed of two membranes situated across the larynx. These membranes modulate the flow of air being expelled from the lungs, being open during inhalation, closed when holding one's breath, and vibrating for speech. Each spoken word is created out of the phonetic combination of vowel and consonant sounds. They also vibrate for singing at 400 times per second, producing sounds rich in variable harmonics.

Being able to communicate through a complex organ that produces sounds associated with a large vocabulary is one of the outstanding attributes that has contributed to man's rise to the pinnacle of life on this planet.

*"As you are reading these words, you are taking part in one of the wonders of the natural world. For you and I belong to a species with a remarkable ability: we can shape events in each other's brains with remarkable precision. . . an ability that is uncontroversially present in every one of us. That ability is language. Simply by making noises with our mouths, we can reliably cause precise new combinations of ideas to arise in each other's minds. The ability comes so naturally that we are apt to forget what a miracle it is."—Steven Pinker, from his book, "The Language Instinct: The New Science of Language and Mind"*

Aside from the ability of a few birds to mimic human speech, no other animal is anatomically equipped for speech—not even members of the ape family, man's closest biological relatives.

Evolutionists have been unable to show how man developed the ability to speak. Human language appears to be a unique phenomenon that appeared suddenly without any transitional phase through other animals. This appears to be another example of a guiding hand assisting in man's development.

The impact of spoken words on the history of man has been profound—whether it was Hitler's demonic exhortations that resulted in the death of millions, Winston Churchill's orations lifting the spirits of his countrymen during World War II, Franklin D. Roosevelt's fireside chats during the Great Depression, the stirring words of John F. Kennedy in his inaugural speech or Martin Luther King's "I Have a Dream" speech—all had the power to impact the minds and emotions of many people.

The ability to speak and sing is woven into the DNA blueprint of man. One has only to listen to a song sung by a choral group, Celine Dion, Andrea Bocelli or other singers to know how bleak life would be without the beauty of these voices that lift our spirits and immerse us in their rapturous sound. To this day, hearing Judy Collins singing "Amazing Grace" or a chorus singing Beethoven's "Ode to Joy" can bring tears to many eyes.

THE BRAIN
The human brain simultaneously processes an incredible amount of information. The brain processes your feelings, thoughts and memories. Simultaneously it also monitors and controls all the functions of your body—muscle movement, breathing, the beating of your heart, hearing,

eyesight, sensory feelings of your skin and a multitude of other functions needed to keep you alive.

It is estimated that the human brain processes more than a million messages a second. The brain must evaluate all this information instantaneously, deciding what is important and what is not. If the brain did not do this our body would be frozen in the decision making process, unable to function effectively in its environment. And most important, one aspect of the power of man's brain, setting him apart all other living animals, is his possession of self-awareness and the ability to analyze complex situations and take appropriate action.

## THE CARDIOVASCULAR SYSTEM

The cardiovascular, or circulatory, system includes the heart, blood vessels and blood. It is vital for fighting diseases and maintaining the proper temperature and pH balance in the blood. The system's main function is to transport blood, nutrients, to and from the cells throughout the body.

Here are some facts about this vital system[1]
- The heart beats about 3 billion times in average person's life.
- About 9 million blood cells die in the human body every second., and the same number are born each second.
- Within a tiny droplet of blood there are some 5 million red blood cells.

- It takes about twenty seconds for a blood cell to circle the whole body.
- Red blood cells make approximately 250.000 round trips of the body before returning to the bone marrow, where they were born, to die.
- Blood cells may live for about four months, circulating throughout the body, feeding the 60 trillion other body cells.

[1] http://warriors.warren.k12.il.us/dburke/amazingfactscirculatory.htm

If you were to lay out all of the arteries, capillaries and veins of one adult, end-to-end, they would stretch out about 60,000 miles. Since circumference of the Earth is about 25,000 miles the vessels could go around the Earth approximately 2 ½ times. Capillaries are tiny, measuring about a tenth of the diameter of a human hair and represent about 80% of the circulatory system.

The blood being transported throughout the circulatory system is a miracle fluid. It delivers oxygen, hormones, nutrients, antibodies and white blood cells to fight disease and infections, platelets for blood clotting, electrolytes to maintain the body's pH and other vital substances to the cells throughout the body. It transports waste products away from these same cells and moves through the liver and kidneys discarding these products and picking up nutrients and fluids. When the blood reaches the lungs, a gas exchange occurs transferring carbon dioxide out of the blood and absorbing oxygen into the blood. The lungs themselves are miraculous organs performing their gas exchange function over an internal surface area close to that of a tennis court—20,000 times a day. The newly oxygenated blood is pumped into the left side of the heart in the pulmonary vein and enters the left atrium. From there it passes through the bicuspid valve, through the ventricle and is pumped out into the body through the aorta by the heart's contraction, to again begin its circulatory journey through the body. As with the other organs previously discussed, the functions to be performed by the heart and the circulatory system are laid out in minute detail by the body's DNA blueprint. From a sensory standpoint, there is no sound more beautiful than the sound of a heartbeat as you lay your head on the chest of a human body—for you are hearing and feeling the subtle drumbeat of life itself.

Having presented many of the facts behind some of the biological marvels that comprise the human body, one might ask, does it seem logical that the miracle of the human body came into existence without a guiding hand? Does it seem logical that it all just kind of fell together by chance? A logical look at the facts dictates otherwise. And beyond the scientific facts, there are the sensory aspects of human life—surely gifts from God given to his ultimate creation.

Think of all the sensory pleasures that we experience—music, rainbows, sunsets and sunrises, the beauty of lakes and oceans, the magic of a crackling campfire or fireplace, the beauty of a full moon, the beauty of a starry sky, the sound of the wind rustling through a forest canopy, the beauty of mountains, the sound of waves crashing on a shore, the warmth of the sun, the wail of a loon, the singing of song birds, the laughter of children, the comfort you feel with a baby in your arms, the warmth of a human body next to you bed, the sexual experience, the smell of flowers, the smell of the forest or the desert after a rain, the feel of breezes against your skin, the smell of fresh baked bread and the taste of wonderful food and drink.

I could go on listing many other wonderful ways that our senses bring pleasure to our lives, but one quote says it all.

*"Life is the ticket to the greatest show on Earth."--Martin Fischer*

# CHAPTER 6

## SOME PHENOMENA TO CONSIDER

In the preceding chapter the miraculous engineering of the human sensory abilities were presented as one of the principal reasons that appear to point to the existence of God. In addition consider the following phenomena—which also appear to point to a guiding hand.

### FIRE
Fire gave man, especially early man, warmth from the cold, light to conquer darkness, heat to cook his food and to ward of his enemies. Consider how the process of fire (oxidation) is set exactly right. If the percentage of oxygen in the atmosphere were set higher, materials would burn faster and hotter, and some materials would actively burn that do not burn in a mixture of 21% oxygen. The Earth would literally be engulfed in flames. If the percentage were much lower or higher it is doubtful that man would have survived.

### OXYGEN AND CARBON DIOXIDE
The percentage of oxygen in the air we breathe is higher than any other known world. The level of oxygen in the atmosphere appears to be in perfect equilibrium today—carbon dioxide is converted to oxygen primarily by plants, and oxygen is, in turn, converted to carbon dioxide by animals and other processes. Although the amount of carbon dioxide in the atmosphere has increased somewhat (2.21 parts per million 2005-2014) it remains to be seen whether small increases such as this will have a deleterious effect on the Earth's climate.

Consider the following:

*"A small sample of the oxygen molecules from any breath that anybody took within the past thousand years is certain to be in the next breath you take. Name a historical figure—Lincoln, John Wilkes Booth, Cleopatra, Hitler, your great grandmother. Tiny samples of them all are in the air you have just drawn in."—David Bodanis, author of "The Secret House: The Extraordinary Science of an Ordinary Day", based on studies that indicate oxygen molecules are in constant motion and can travel up to 1,000 miles in as little as two weeks.*

## WATER

Water is colorless, odorless and without taste and no living thing can survive without it. Approximately two-thirds of the human body is water. The bodies of other animals and plants also consist mostly of water.

The attributes of water are particularly aimed to support life—it has a high boiling point and freezing point when compared to other liquids. Water is a universal solvent and chemically neutral, which means that many minerals, chemicals, and nutrients can be carried throughout the human body. Because of its properties water is an essential component of DNA—and appears to be the perfect liquid for the existence of life. In addition, because of water's surface tension it is able to flow upward against gravity within a plant's vascular system, carrying nutrients to the tops of the tallest trees.

It is fortuitous that there is an ample supply of water in many areas of the world. Consider the following:

*"... 70 per cent of the world's fresh water is used for agriculture ... one egg requires 120 gallons to produce... 100 gallons for a watermelon . . . meat is among our thirstiest commodities, requiring 2500 gallons per pound."—From "Abundance" by Peter Diamandis and Steven Kotler*

Ninety-seven percent of the Earth's water is in the oceans and unusable by man. However, there is a miraculous system of recycling and purification in place. Evaporation takes the ocean waters into the atmosphere in fresh water form, leaving the salt behind. Earth's meteorological system then forms clouds, which moved by winds, caused by the Earth's rotation, distribute the water throughout the world for the use by plants, animals and people.

Every molecule of water on Earth has been here since the planet formed, over four billion years ago. The water you drink today has been recycled countless times through the evaporation-rain cycle, flowed through rivers, cascaded over falls, floated down to the Earth as snowflakes, crashed against ocean shores, been taken from the soil by the roots of plants, billowed through the sky as clouds and moved through the digestive systems of insects, fish, mammals and dinosaurs.

Surface albedo, the reflective property of the Earth's atmosphere, is set just right, principally because of its cloud cover. If it were too high or too low the Earth would either be a frozen ball or a hot, dry planet.

## METEORS

As discussed earlier, a meteor, striking in the Gulf of Mexico, off the coast of Mexico, 65 million years ago, ended the Mesozoic Era of the dinosaurs, resulting in the eventual rise of man. Much of the debris entering or circulating within our solar system is "vacuumed up" by the larger planets, Jupiter and Saturn, due to their gravitational pull. However, some of these wandering pieces of space debris cross Earth's path.

A very rare recent event occurred on February 15, 2013, when a large 10,000 ton meteor traveling at 43,000 miles per hour, exploded over Chelyabinsk, Russia, injuring 1500 people. Miraculously, no one was killed. Coincidentally, on that same day, an asteroid estimated at 40,000 tons, passed within 17,200 miles of the Earth's surface, closer than many of the satellites above the Earth—also a very rare event. Had it struck the Earth it would have caused catastrophic damage. The probability of both these events happening on the same day is astronomical (no pun intended). Perhaps someone was giving us a message.

*"Hold your hand out to catch a few raindrops, and a particle, older than the Earth, which has traveled trillions of miles, and just arrived from outer space a few weeks before, will be in your hand."—David Bodanis, from his book "The Secret House—The Extraordinary Science Of An Ordinary Day", pointing out that the Earth is bombarded with over 6,000 tons of micro-meteorites from outer space each day, which float around in the upper atmosphere before many of them fall to Earth in raindrops.*

## PHOTOSYNTHESIS

The miracle of photosynthesis occurring in the green leaves of a plant produces food for the plant and oxygen for the atmosphere. It is a very complicated process—one that is spelled out in the DNA of the plant—again, a blueprint telling the organism how to conduct this intricate process. One of the reactions in the chain of chemical processes that comprise photosynthesis must take place in 3 trillionths of a second or the process would fail. Without the photosynthesis process there would be no life, as we know it, on Earth.

## NITROGEN FIXATION

Nitrogen, which comprises approximately 78% of the atmosphere, and is essential to a plant's growth, must be processed, or "fixed" in order to be used by plants. Most fixation is done by bacteria in the soil, which have an enzyme that combines gaseous nitrogen with hydrogen to produce ammonia, which is subsequently converted by the bacteria into other organic compounds. Nitrogen-fixing bacteria live on the roots of legumes such as peas, string beans and alfalfa, where they form a relationship with the plant, producing ammonia in exchange for carbohydrates. Because of this relationship, legumes increase the nitrogen content of nitrogen-poor soils. Due to the ability of such crops to increase the nitrogen content of soil, they are extensively planted in alternate years as a crop rotation measure. Consider how important this process is to the life of plants—truly a miracle being performed by living bacteria in the soil—bacteria programmed with a DNA blueprint directing them how to perform this activity.

# THE SIZE, LOCATION, ROTATION AND AXIS OF THE EARTH

Consider the following facts about the Earth we inhabit:

- The size of the Earth is ideal—if it were larger gravity would be too strong to allow life to exist as we know it today; if it were smaller, the atmosphere would dissipate into space, producing an airless, desolate planet. There are examples of both of these extremes in our solar system.

- The Earth is located in the "Goldilocks" zone of our solar system—a small change in the distance from the sun would have resulted in either a hot, arid planet or an icy ball. The sun's location in the outer arm of our Milky Way galaxy is also in a "Goldilocks" zone, since if it were much closer to the center, the Earth would be subject to increased impacts by comets and radiation from exploding stars.

- The rotation of the Earth is set to allow for the existence of a livable planet. If slower, it would have extreme temperatures; if faster, the atmosphere would produce violent winds, and extreme weather events. How the Earth attained this perfect time of rotation is an unanswered question.

- The combination of the Earth's rotation and the inner solid core produces an electromagnetic field, which protects the Earth from the solar wind and cosmic rays that would strip away the Earth's ozone layer and expose the Earth to harmful ultraviolet radiation, killing much of its life, including humans.

- The axis of the Earth allows the progression of seasons, and mitigates any extremes of temperature over much of the planet. Without the axis, which appears to be set at an ideal amount, the area close to the equator would be almost unlivable due to the heat and the southern and northern parts above the equator would have a very small habitable zone, with extremely low temperatures at the pole regions—far lower than today. Astronomical investigation has determined that the Earth was knocked onto its present axis by a collision with a Mars size planet, early in the solar system's formation—an extremely fortuitous event.

## YOUR PLACE IN THE UNIVERSE

As you sit quietly reading this book you might consider the following:

*"The next time someone in your family or group of friends calls you lazy for just sitting there, you can politely remark that, although it may look as if you are just sitting, you are actually moving at great speed around the Earth (1000 miles an hour at the equator), around the Sun (66,000 miles per hour), through the Milky Way (43,000 miles per hour), around the arm of our galaxy (483,000 miles per hour) and in our galaxy through the universe (1,300,000 miles per hour).—Andrew Fraknoi.*

You could add six more movements to the five indicated above if you were sitting in a conveyance like a boat or plane—three for the movement of the conveyance (up and down, side to side, back and forth {heave, sway and surge}) and three for the attitude around the center of mass of the conveyance (roll, yaw, pitch).

Seems you aren't just sitting there after all

## THE SUN
The sun at the center of our solar system is the source of energy for life on Earth. It is composed principally of the elements hydrogen and helium with small amounts of the heavier elements, oxygen, carbon, iron, neon and a few others. The source of its tremendous energy is fusion, a complicated process occurring in the sun's core that basically converts two atoms of hydrogen into one atom of helium, annihilating an amount of matter in the process.

*"The bomb over Japan destroyed an entire city, simply from sucking several ounces of uranium out of existence, and transforming it into glowing energy. The reason the sun is so much more powerful is that it pumps 4 million tons of hydrogen into pure energy every second."—David Bodanis, from his book "$E=mc^2$".*

Lest the reader feel that the sun will shortly burn itself up, since its hydrogen fuel is disappearing at such a rate, it is estimated that at the present rate of converting hydrogen to helium the sun will maintain its current state of energy production for another 5 billion years.

## THE MOON
Roughly a quarter of the size of the Earth, the moon is believed to have been formed from debris that broke off from the Earth early during its formation about 4 ½ billion years ago by a collision with a Mar's size planet. Originally, the moon was much closer to the Earth. The moon's current location, approximately 240,000 miles

from Earth provides an ideal gravitational pull upon the Earth's oceans, causing tides along its shores that create special areas for aquatic life to live and propagate.

One of the most important impacts of the moon's gravitational pull on the Earth is its stabilizing effect upon the Earth's rotation. If the rotation were not stabilized there would be very strong winds and erratic weather patterns over the Earth's surface.

TOTAL SOLAR ECLIPSES
A total solar eclipse is an astronomical event that struck awe into the hearts of early man. The sight of the sun's disc disappearing behind the disc of the moon was thought to foretell tragic future events. Today it is viewed as a stunning visual event for those lucky enough to be in its path on a cloudless day. It is a sight that can't be duplicated on film or by photo. It has to be experienced in person.

*"Eclipses are different (than other solar events). They truly fit the "amazing coincidence or divine plan" category. How else to explain that the moon is four hundred times smaller than the Sun but also four hundred times nearer to us? This makes the only two disks in our sky appear the same size. That would not be the case if either were larger, smaller, nearer, or farther away. Even more in line with "a divine plan" is the fact that since its creation four billion years ago from an interplanetary collision the moon has been slowly receding from the Earth at the rate of 1- ½ inches a year. During the time of man's evolution on Earth the moon's face is currently in an optimum position that matches the face of the sun."—Bob Berman from his book "The Sun's Heartbeat".*

People living in the United States will be in an enviable position on April 8, 2024, when a total solar eclipse will cross the North American continent, passing near many major urban centers. There is one point, near southern Illinois, which was on the path of an earlier eclipse that took place on August 21, 2017. It's very rare that any specific point on Earth would be on the path of a total eclipse of the sun twice within seven years. To give some perspective to this probability, the next solar eclipse in Los Angeles will occur on April 1, 3290, 1,566 years since the last eclipse on May 22, 1724.

THE SINGING OF BIRDS
Think how lonely are lives would be without the singing of birds that are an integral part of the cities, forests and fields. Their melodious songs and chirping add a dimension to a spring day that is difficult to describe. One of the loveliest sounds for those fortunate enough to have heard it is the call of the loon on a quiet evening on a lake in the north woods.

*"The loons were calling, I can hear them yet, echoes rolling back from the shores and from unknown lakes across the ridges until the dusk seemed alive with their music."—Sigurd Olson, naturalist.*

To hear the wail of a loon during a quiet evening is an experience that pierces you to your soul—a sound never to be forgotten.

These are only some of the phenomena that we encounter during our life here on Earth. Once again, we have to ask—Is this all by chance, without the guiding hand of God? Did it all just kind of happen? Sit back, close your eyes and ponder this.

One last thought to consider—what if you were God? What kind of environment would you have created? Could you have improved on God's work? Would the Earth have been much different than the wonderful one that exists for us today? Does it seem likely it all just came together without a guiding hand? Consider all the laws of existence and the events that had to take place to provide the environment we have today — breathable air, photosynthesis, nitrogen fixation, the axis of the Earth and its rotation, the effect of the moon upon the seas of the Earth and the Earth's rotation, controllable oxidation (fire), the qualities of water, the position of the Earth in the "Goldilocks" zone from the sun, the gravitational pull of the larger planets to ward off and absorb many of the asteroids and comets that exist within the solar system, the gravitational force that is neither too strong or too weak, and many others too numerous to list. And then think of the visual sights and sounds of nature that God gave us to appreciate with the miraculous sensory abilities he instilled in us through the programs of our DNA—rainbows, sunrises, sunsets, waterfalls, ocean waves pounding the shores, fields of flowers, songs of birds, crackling campfires, starry skies, full moons, and coral reefs teeming with aquatic life.

One has to wonder if God isn't giving us a glimpse of what heaven is like.

# CHAPTER 7

## ALONE IN THE UNIVERSE?

There are some scientists who believe that the universe contains many Earthlike inhabitable planets. However, remarkably, the number of scientists who believe this has been shrinking over time as more evidence arises that indicates that the conditions which allowed life here on Earth to proliferate appear to be less likely to exist in other parts of the universe

In 1950 Physicist Enrico Fermi, asked his famous and often quoted question, "Where is everybody?", implying that it was highly improbable that extraterrestrials had not visited the Earth, given the age of the universe. One possible explanation put forth by John Gribbin in his book, "Alone In The Universe", is that the conditions and events that led to the formation of the Earth and the life upon it are so unique and improbable that we are, in fact—alone in the universe. If one accepts that God created man and the planet on which he dwells it is very probable that God has made the Earth and its inhabitants unique, leading to the conclusion there is no life anywhere else in the universe—certainly no life as we know it

At one time, in the 1970s, it was believed there were only a couple of requirements to produce a viable Earthlike condition on a planet which would support life—a comparatively small rocky body circling a sun at a correct distance—in the so-called "Goldilocks" zone. Under those requirements it was estimated there would be thousands, if not hundreds of thousands, of Earthlike planets in our Milky Way galaxy alone.

As time has gone by the number of requirements has increased, and the estimate of potential Earthlike bodies has decreased substantially. Today, it is estimated there are more than 200 requirements to produce an Earth like ours that could support life. For example, as noted before, the Earth is located the at just the right distance from the sun. Consider the temperature swings we encounter, roughly -30 degrees to +120 degrees. If the Earth were any further away from the sun, we would all freeze. Any closer and we would burn up. Even a fractional variance in the Earth's position to the sun would make life on Earth impossible. The Earth remains this perfect distance from the sun while it rotates around the sun at a speed over 66,000 mph. Other planets have rotational paths that vary widely. The Earth is also rotating on its axis, allowing the entire surface of the Earth to be properly warmed and cooled every day. Scientists are still not sure how the rotation of the Earth came to be set at this optimum level.

A planetary collision with a Mars-size object, which produced our moon, appears to also have knocked the Earth onto an axis which produces seasons. Without that tilt, the development of life on Earth, as we know it, would have been much different. In addition, the existence of Jupiter's strong gravitational pull serves as a magnet, preventing most asteroids and other potentially destructive bodies from striking the Earth.

These are only a few examples of the 200 plus requirements. The more research goes forward in this area the more likely the number of requirements will increase.

The Search For Extraterrestrial Intelligence (SETI), an expensive, privately funded project, once partially funded by the federal government, begun in the 1960s, has searched the heavens in vain with its vast radio telescopic network, listening for any evidence that would herald the existence of other intelligent life in our galaxy. Recently, one of its leading spokesmen had this to say:

*"In light of new findings and insights, it seems appropriate to put excessive euphoria to rest . . . We should quietly admit that the early estimates . . . may no longer be tenable."—Peter Shenkel*

It appears that the fortuitous outcome of the receipt of messages from another civilization portrayed in the 1997 movie *Contact* is merely wishful thinking given the multitude of requirements for an advanced civilization and the vast distances involved.

Today many researchers are becoming dubious that planet such as ours exist anywhere else. In fact, some scientists are amazed that a life sustaining planet such as our Earth exists at all, given the tremendous odds against its existence.

# CHAPTER 8

## THE DEVIL'S ADVOCATE

At this point it would be worthwhile to lay out the position of the Devil's Advocate. In the Roman Catholic Church a Devil's Advocate is assigned to present evidence against a case for the pronouncement of sainthood for a person. In this case the arguments against the existence of God will be presented.

One of the strongest arguments (or hypothesis) for the creation of a universe without God is this: Since the physical laws of the universe are so perfectly set a non-deity force caused a vast multitude of universes to randomly form—the Multiverse Hypothesis. Other innumerable variations of universes with different physical laws were unstable and did not survive—we are in the lucky one that did. Some could question the amount of time involved since the potential combination of physical laws governing a universe's existence are almost beyond comprehension. However, we are speaking in terms of eternity so the creation of these innumerable universes might be accomplished. It's similar to the hypothetical situation described in which millions of monkeys are set before typewriters and start banging away. Their output, almost its entirety, would be gibberish. However, at some point, since we are working with eternity, one of the monkeys would type out a complete set of Shakespeare's plays. What goes unanswered is who or what is behind the curtain on this magical machine that produces these universes.

There is a philosophical concept titled Occam's Razor (Ockham's Razor), which dates back to the 1300s. William of Ockham, an English Franciscan friar, stated, in essence, that in attempting to explain some problem or event it is the simplest explanation that is probably the most accurate. The concept of a God that created the universe seems an easily understood simple answer to the question as to why an ideal universe exists. The complex, convoluted and problematical concept of a universe-making machine seems to be a futile grasp at straws by scientists to come up with a "universe creator" in order to deny God's existence. There is absolutely no scientific evidence supporting this hypothesis. And finally, even the multiverse concept needs a beginning.

There are also the problems that appear in the investigation of DNA and the structure of our own human body. There are many "dead ends" in the DNA computer code that appear to go nowhere, as far as we can determine. Although as our knowledge increases we find that some parts originally thought as useless appear to have purpose. With regard to the human body, biologists point out that the melding of the human tubes for eating and breathing would have been better structured if they were separated. Numerous choking problems occur when food enters our breathing system.

The other arguments against the existence of God enter the philosophical realm. Why does God allow horrible things to happen to good and innocent people? Millions died in the plagues of the Middle Ages and influenza epidemics in the 20th century. Even today we have the

Ebola and Coronavirus epidemics that cause many deaths.. There are literally thousands of illnesses and ailments that wreak havoc on the human body—cancer being one of the worst. There are also the natural tragedies such as earthquakes, tornados, tsunamis and hurricanes that cause countless deaths. Surely a merciful God, if he existed, would not allow these things to happen to innocent people.

We also have the man-made tragedies such as the Holocaust and world wars. What kind of God would allow things like this to occur? Surely the prayers of the poor souls swept up in these terrible events would cause his intervention if he existed.

The occurrence of man-made tragedies, human ailments and natural catastrophes is, in my mind, the strongest argument against God's existence that atheists can put forth. Atheists stress these instances in denying God's existence—but that does not disprove the scientific evidence presented in the preceding chapters. Proving God's existence is one thing—not comprehending the vicissitudes and hardships of life on this earth is another.

Another argument against the existence of God is the very act of the creation itself. Why did God feel the need to do it? Did God feel something was missing in his happiness and thus had the urge to create? Sounds more like a bored person here on Earth. And what's the purpose of creating an entire universe of innumerable galaxies and stars? Surely just the Earth and the sun would have been sufficient. It's difficult to believe in a God who is all-powerful, all-knowing and completely without any of the human needs that we experience, possessing a need to create the universe and intelligent life. However, one response to this line of criticism is that God created the universe not for his good but for ours.

Discussed above are the major arguments against God's existence. Some of them can never be answered to the satisfaction of some people—especially those attempting to understand the mind of God. However, an attempt to provide an answer will be made in the next chapter.

# CHAPTER 9

## WHAT KIND OF GOD?

After coming to the conclusion that there is a God, based on the evidence presented in the preceding chapters, I asked myself, what kind of God is it? I have tried to formulate an answer. What follows from this point forward is pure speculation.

It's hard for me to accept a God who gets involved in the daily activities of people. I believe he has given us free will and pretty much lets us do as we please. That would explain how Hitler, Stalin, Mao and the other tyrants of history caused the horror and suffering they did. It also explains why the pleas of people in concentration death camps are ignored, and prayers go unanswered in so many cases. When prayers are answered, God is thanked; when they go unanswered we're told, "It's God's will." Doesn't make a lot of sense to me. I believe man is the captain of his ship, the master of his soul, and that humans are responsible for what happens to other humans.

It is difficult to square the concept of an interventionist God on the individual level with the facts of life we see before us, such as deformed babies, people born into scarcity, disease and poverty, innocents caught up in war, extermination camps, genocide, etc.

There are other facts of life here on Earth that are very troubling for me. If we allow for the possibility of a judgment day and life after death how do we explain those people born with diminished mental capacity such as Down's syndrome, schizophrenia and advanced autism. In addition, we have those people who are affected by traumatic injuries to the brain, and those affected by Alzheimer's—two conditions which diminish a person's mental capacity, and ultimately their ability to distinguish right from wrong—or even their awareness of being. And what of babies or young children who die in car accidents, fires or airplane crashes? And what of the babies yet unborn that are aborted? Do they have souls? How are these lives to be judged? These are situations for which I have no answer. Perhaps these souls are recycled into new beings born in the future—a form of reincarnation. I just don't have an answer to these questions.

I find it difficult to imagine God as described in the Old Testament. It's hard to believe the entire Bible was divinely inspired, because the Old Testament God is surely not someone who is "Godly"—he is a mean, vindictive God, whose human qualities of anger and revenge were invented, I believe, by the writers of the time. Surely God is able to rise above these base qualities. I believe in a loving God, who appreciates expressions of thanks for the gift of life he has given us, not a God who has to be praised all the time. In fact,

I see him more as a valued friend who I can speak to throughout my life and thank daily for the beauty of the world he has created for me. I also believe that Jesus did exist and that what he preached has relevance for all men—for all time

The universe having been set into motion by God, and bound by the physical laws described in the previous chapters, phenomena like tornadoes, hurricanes, lightning strikes, volcanoes, earthquakes and floods, can, and do, cause death and destruction in a haphazard manner—including the death of young innocents praying in churches. Our earthly environment, as beautiful as it is, was never meant to be heaven. There are serious risks surviving the environment that God has created. In a nutshell, I believe God is saying, "Tomorrow is promised to no one."

Since we can't look into a telescope and see God sitting on a throne in heaven much of what an individual believes about God and the afterlife must be based on faith.

The American philosopher William James is quoted as saying that religious experience emanates from 'feeling at home in the universe.' Since everything on Earth, including our bodies, is composed of "star dust" from the death of stars, we are truly at one with the universe in which we live.

I believe in a God who loves us and is disappointed when we fall short of leading a good life and treating all living things with compassion.

And although the reason for our existence cannot be known I choose to believe in the all-encompassing statement made by Rick Warren, church founder and author of "A Purpose Driven Life."—

*"If you're alive, there's a purpose for your life. You were made by God and for God, and until you understand that, life will never make sense."*

There are many organized religions here on Earth. Who is to determine which one is the right one? And what of the people who lived before today's great religions, espousing a single omnipotent God? Have they no place in heaven? In the end we must all make our own choices, both in our mind and in our heart. Those religions which consider themselves the one and only true path to God, cannot, in my view, have the approval of God. Since we are greatly influenced by the environment in which we were raised our religious choice is already made for us in many cases. In that part of the world under Communist rule that choice in many cases is atheism.

In the end, I find that my view matches most closely the words spoken by an East Indian religious figure, Indra Devi . . ."*I do not belong to any religion. Everything is between God and myself.*"

I believe, without proof, in an afterlife and a day of judgment for the life one has lived. I believe there should be some punishment for those who made choices that hurt others but I find it difficult to imagine a punishment of eternity in hell.

A belief in an afterlife could just be wishful thinking. Heaven, is an unknowable concept to me. I have truly found heaven here on Earth many times, so a heaven in which my spirit wanders in the ethos for eternity is an incomprehensible concept for me.

However, since God created the universe and man it seems illogical he would let us dance across the stage of life and into oblivion. He would seem to have a higher purpose in mind. Since I believe he is a loving God it seems to me he would want to bring happiness to those who he brought into existence and endowed with souls.

*"Surely God would not have created such a being as man, with an ability to grasp the infinite, to exist only for a day! No, no, man was made for immortality."—Abraham Lincoln*

Having come to the conclusions described above I find that there are one important question for which I have no answer—What is God's ultimate plan?

In the end we are left with the thought of one famous writer regarding these questions.

*"A God that can be understood is no God. Who can explain the Infinite in words?"—Somerset Maugham*

## CHAPTER 10

## WRAPPING IT ALL UP

The preceding chapters laid out the evidence for the existence of God. This evidence does not support a God who intervenes in the daily activities of humans but does support his involvement in events that shaped the world we have today

The existence of God explains many things:
- Why there is something instead of nothing.
- Why there are physical laws that hold the universe together that have been exquisitely set.
- Why life exists.
- Why there are DNA codes for all living things—from one celled organisms to human beings.
- Why the Big Bang occurred and its explosive force was set so perfectly—within one quadrillionth of one percent
- Why the rotation and the axis of the Earth have been set so perfectly, allowing temperate climates over much of its surface.
- Why the moon is at the exact distance from the Earth so that its face exactly covers the sun in a total solar eclipse—a stellar astronomical event.
- Why an extinction event occurred 65 million years ago clearing the path for man's existence and future domination of the planet—an example of another dramatic event set at exactly the right

magnitude, killing all the dinosaurs but leaving small mammals to evolve.
- Why music, truly a gift from God, exists.
- Why man, among all living things, is endowed with the power of speech.
- Why the oxygen level is set so perfectly to allow fire at a controllable level and for man to exist.
- Why the incredibly complicated process of photosynthesis exists to generate food and oxygen, for both plants and animals.
- Why the process of nitrogen fixation exists.
- Why water, essential for the existence of all living things, is abundant, being generated by the process of evaporation from the salt-laden ocean waters and distributed as rain or snow in freshwater form throughout the globe.
- Why the Earth generates an electrical field that deflects harmful cosmic rays from the sun due to its solid iron core and its perfectly set rotation.
- Why the beauty of the Earth's environment exists.

There is other evidence for God's existence but I believe I have listed the major ones. I believe the facts above,, standing alone, prove God's existence. Although any one fact many not convince a person it is the accumulation of them all. They all appear to point in one direction. I believe I would be a fool to believe otherwise.

# CHAPTER 10

## FINAL THOUGHTS

Being an agnostic or an atheist years ago was much easier than it is today. The scientific advances which uncovered the Big Bang, the physical laws holding the universe together, the DNA blueprint, and the power within the atom have made the case for God's existence very strong.

I see the shadows slowly lengthening in my life. What awaits me is the last great event that everyone will experience. Some fear that event; others are reluctantly resigned to it. And others, like me, see it as an essential part of the tapestry of life. I want to live as long as I have my physical and mental health, but when the time comes that my quality of life deteriorates I want to take that walk towards the bright light that some have experienced in near-death situations. That bright light may merely be the sensory explosion resulting from the death of one's body—or it may be the portal into the afterlife. What lies beyond that bright light cannot be known by us. I do know, however, that there is a God and this book has been written to state the reasons I came to that conclusion after much thought.

The choice was simple:

- Either the universe spontaneously sprang into being from nothing with all the physical laws and DNA blueprints set exactly right for our existence by chance, or ...

- The prime mover for the creation of the universe was an external force, the first cause, an all-powerful supreme being—God.

I chose the second alternative. I cannot say that God will grant me an afterlife, but I do believe there is a God and I thank him every day for the gift of life he has given me.

## RELEVANT QUOTES

"The more I study science, the more I believe in God."—Albert Einstein

"The most miraculous thing is happening. . . .The physicists are getting down to the nitty-gritty; they've just about pared things down to the ultimate details, and the last thing they ever expected to happen is happening. God is showing through. They hate it, but they can't do anything about it. Facts are facts.—John Updike

"Scientists are slowly waking up to an inconvenient truth - the universe looks suspiciously like a fix. The issue concerns the very laws of nature themselves. For 40 years, physicists and cosmologists have been quietly collecting examples of all too convenient "coincidences" and special features in the underlying laws of the universe that seem to be necessary in order for life, and hence conscious beings, to exist. Change any one of them and the consequences would be lethal."—Paul Davies, physicist

"The existence of design and nature is a fact which must certainly be taken seriously ... (because) in every main branch of science- physics, geophysics, astronomy, chemistry, biology- we are faced by the same surprising fact.... Nearly everywhere it (nature) shows the signs.... of something that we can only think of in terms of ingenuity and deliberate design."—Robert E. D. Clark, PhD, Organic Chemistry, Cambridge University

"Intelligence is responsible for the sequencing of the nucleotides (DNA)"—Hubert P. Yockey, physicist and information theorist

"Perhaps God wants to be known not by everyone but only by the creatures who seek Him."—Blaise Pascal

"For the (atheistic) scientist who has lived by his faith in the power of reason, the story ends like a bad dream. He has scaled the mountains of ignorance; he is about to conquer the highest peak; as he pulls himself over the final rock, he is greeted by a band of theologians who have been sitting there for centuries." —Astronomer Robert Jastrow, commenting on the discovery that the universe was created in a "Big Bang" 13.8 billion years ago—from nothing, with the only possible answer being creation by a supernatural force.

"A commonsense interpretation of the facts suggests that a superintellect has monkeyed with physics, as well as chemistry and biology, and that there are no blind forces worth speaking about in nature. . . The numbers one calculates from the facts seem to me so overwhelming as to put this conclusion almost beyond question." —Physicist Fred Hoyle, commenting on the conclusion he reached after studying the complexities of life, the "Big Bang" and the many laws of physics which are tuned "just right" to allow for the existence of the universe.

"When I first open my eyes upon the morning meadows and look out upon the beautiful world, I thank God I am alive."—Ralph Waldo Emerson

"The (human) genome is a scripture in which is written the past history of plagues. The long struggles of our ancestors with malaria and dysentery are recorded in the patterns of human genetic variation. Your chances of avoiding death from malaria are pre-programmed in your genes, and in the genes of the malaria organism. You send out your team of genes to play the match, and so does the malaria parasite. If their attackers are better than your defenders, they win. Bad luck. No substitutes allowed."—Matt Ridley, from his book *Genome: The Autobiography Of A Species In 23 Chapters*

"Never let religion get between you and your God. Man created religion, not God."—Park Brady in a Letters to the Editor in *The Wall Street Journal*, in a response to the recent Roman Catholic Church scandal involving pedophilia.

"Belief in God and in immortality gives us the moral strength and the ethical guidance we need for virtually every action in our daily lives."—Rocket Scientist Wernher von Braun

"When I stand before God at the end of my life, I would hope that I would not have a single bit of talent left, and could say, 'I used everything you gave me'."—Erma Bombeck

"We thank God for our homes and our food and our safety in a new land. We thank God for the opportunity to create a new world for freedom and justice." — Elder William Brewster, minister, at the first Pilgrim feast in the New World, considered the first Thanksgiving.

"When you hear music that calls out to you it will tell you something about yourself. Once you enter its domain and it enters yours, it will become a life partner, always illuminating who you are and who you were. It will be beautiful because it is yours."—John Mauceri, from his book *For The Love Of Music: A Conductor's Guide To The Art Of Listening*

"I believe God speaks to us through music."—Grace Botello, after hearing the 1st movement of Felix Mendelssohn's Violin Concerto.

"Pray as if everything depended on God and work as if everything depended on man."—Saint Augustine

"You can't be angry with God and not believe in him at the same time."—Sara Cooper

"I would rather live my life as if there is a God and die to find out there isn't, than live my life as if there isn't and die to find out there is."—Albert Camus

"With a good conscience our only sure reward, with history the final judge of our deeds, let us go forth to lead the land we love, asking His blessing and His help, but knowing that here on Earth, God's work must surely be our own."—President John F. Kennedy, closing words of his inaugural address

"If the stars should appear just one night in a thousand years, how would men believe and adore!"—Ralph Waldo Emerson

"Sometimes it seems like God is difficult to find and impossibly far away. We get so caught up in our small daily duties and irritations that they become the only things that we can focus on. What we forget is that God's love and beauty are all around us every day, if only we take the time to look up and see them."—William Mathias

"Oh, wow. Oh, wow. Oh, wow."—dying words of Steve Jobs, founder of Apple. One has to wonder what he experienced at that time.

"Leave God out of it. Men make war, not God."—Response from an elderly man to a woman who had said, "If there was a God He would have shown some mercy to them.", after the devastating bomb attack on Hamburg, Germany, during World War II as she watched bodies being stacked into trucks. From The American Heritage Picture History of World War II.

"It can be cold. You can get lonely. You can be hurt and sore. These are just challenges. There are challenges every day. Any woods can beat you no matter how good of a woodsman you are. I haven't lost my respect and awe for what God has placed out there for me to see and appreciate. Never." — M. J. "Eb" Eberhart, age 70, 1,400 miles and 46 days into his effort to hike the 4,400 mile North Country Trail, as related to Rick Olivo of The Daily Press, Ashland, Wisconsin.

"What God intended for you goes far beyond anything you can imagine."—Oprah Winfrey

***Trees***

> I think that I shall never see
> A poem lovely as a tree.
> A tree whose hungry mouth is prest
> Against the Earth's sweet flowing breast;
> A tree that looks at God all day,
> And lifts her leafy arms to pray;
> A tree that may in summer wear
> A nest of robins in her hair;
> Upon whose bosom snow has lain;
> Who intimately lives with rain.
> Poems are made by fools like me,
> But only God can make a tree.

—Joyce Kilmer, killed on the Western Front, July 30, 1918.

"God gave you a gift of 86,400 seconds today. Have you used one to say 'thank you'."—William A. Ward

"For anyone who is alone, without God and without a master, the weight of days is dreadful."—Albert Camus

"The essence of life is statistical improbability on a colossal scale."—Richard Dawkins, atheist, from his book "The Blind Watchmaker"

"The belief that there are other life forms in the universe is a matter of faith. There is not a single shred of evidence for any other life forms, and in forty years of searching, none has been discovered. There is absolutely no evidentiary reason to maintain this belief."—Michael Crichton

"It is a travesty of true religion to consider one's own religion as superior and others' inferior. All religions enjoy the worship of one God who is all-pervasive. He is present even in a droplet of water or a tiny speck of dust."—Mohandas "Mahatma" Ghandi

"When I stand on the open prairie early in the morning, just as dawn is breaking, I feel closer to God than anywhere else in the world."—From a rancher's diary. One can also experience this feeling seeing the first rays of the sun break the horizon from a boat or from the shore of a lake or ocean.

"Faith moves mountains, but you have to keep pushing while you are praying."—Mason Cooley

"I can't believe God put us on this Earth to be ordinary."—Lou Holtz

### ***Rainbow Bridge***

Just this side of heaven is a place called Rainbow Bridge.

When an animal dies that has been especially close to someone here, that pet goes to Rainbow Bridge. There are meadows and hills for all of our special friends so they can run and play together. There is plenty of food, water and sunshine and our friends are warm and comfortable.

All the animals who had been ill and old are restored to health and vigor. Those who were hurt or maimed are made whole and strong again, just as we remember them in our dreams of days and times gone by. The animals are happy and content, except for one small thing; they each miss someone very special to them, who had to be left behind.

They all run and play together, but the day comes when one suddenly stops and looks into the distance. His bright eyes are intent. His eager body quivers. Suddenly he begins to run from the group, flying over the green grass, his legs carrying him faster and faster.

You have been spotted, and when you and your special friend finally meet, you cling together in joyous reunion, never to be parted again. The happy kisses rain upon your face; your hands again caress the beloved head, and you look once more into the trusting eyes of your pet, so long gone from your life but never absent from your heart.

Then you cross Rainbow Bridge together.
--Unknown

"A miracle in music and one of those freaks Nature causes to be born."—Pronouncement in 1771 on the musical ability of Mozart by the Italian Academy of Music. More likely Mozart was a gift from God to the human race.

"In every walk with nature one receives far more than he seeks."—John Muir

"I believe in Spinoza's God, who reveals himself in the orderly harmony of what exists, not in a God who concerns himself with the fates and actions of human beings."—Albert Einstein. (Spinoza was a major 17th century philosopher)

"Allahu Akbar (God is great)"—Arabic phrase uttered by fanatical, bomb-carrying Islamic terrorists prior to blowing themselves up along with innocent civilians—usually Muslims like themselves. Their belief that 72 virgins await them for their heinous acts is beyond comprehension.

"Those who can make you believe absurdities can make you commit atrocities."—Voltaire

"We share the air with the forests and the water with the seas. As a body they and we are one."—Buddha

"It is necessary to arrive at a prime mover, put in motion by no other: and this everyone understands to be God."—St. Thomas Aquinas

"Suppose I pitched my foot up against a stone and were asked how the stone came to be there. I might possibly answer . . . it had lain there forever. But suppose I found a watch upon the ground. I should hardly think of the answer I gave before." —Anglican theologian William Paley, in his book Natural Theology, published in 1892, arguing for an Intelligent Designer as the most probable explanation of the complexity of life and the world around us.

"God is or is not. A game is being played and the coin will come down heads or tails. How will you wager? Let us weigh up the gain and loss involved in calling heads that God exists. If you win you win everything. If you lose, you lose nothing. Do not hesitate, then: wager that he does exist."—Blaise Pascal

"As a human being, one has been endowed with just enough intelligence to be able to see clearly how utterly inadequate that intelligence is when confronted by what exists."—Albert Einstein

"The seed of everything that has happened in the universe was planted in that first instant; every star, every planet and every living creature in the Universe came into being as a result of events that were set in motion in the moment of the cosmic explosion...The Universe flashed into being, and we cannot find out what caused that to happen."—Astrophysicist Robert Jastrow, describing the moment of creation known as the Big Bang—14 billion years ago.

"I myself believe that the existence of God lies primarily in inner personal experiences." — William James

### ***High Flight***

"Oh, I have slipped the surly bonds of Earth,
 And danced the skies on laughter silvered wings.
 And while with silent lifting mind I've trod,
 The high untrespassed sanctity of space,
 Put out my hand and touched the face of God."
—John G. Magee  This poem was read to the nation by President Ronald Reagan following the tragedy of the Challenger disaster in January 1985.


"I do not feel obliged to believe that the same God who has endowed us with sense, reason, and intellect has intended us to forgo their use."—Galileo Galilei


"Some people think broadcasting into the universe is like shouting in a jungle—not necessarily a good idea."—Seth Shostak, senior astronomer at the SETI Institute, commenting on the concern of some astronomers that purposefully sending messages into outer space seeking to make contact with aliens could attract those who aren't necessarily benign
.

"Down on the lake rosy reflections of celestial vapor appeared, and I said, 'God, I love you' and looked to the sky and really meant it. 'I have fallen in love with you, God. Take care of us all, one way or the other.' To the children and the innocent it's all the same."—Jack Kerouac, "The Dharma Bums"

"Prayer does not change God, but it changes him who prays."—Soren Kierkegaard

"I look at the universe and I know there's an architect."—Jack Anderson

"If the only prayer you ever say in your entire life is thank you, it will be enough."—Meister Eckhart

"Mozart makes you believe in God, because it cannot be by chance that such a phenomenon arrives into this world and leaves such an unbounded number of unparalleled masterpieces."—George Solti

"I cannot imagine a God who rewards and punishes the objects of his creation and is but a reflection of human frailty."—Albert Einstein

"Never lose an opportunity of seeing anything beautiful, for beauty is God's handwriting."—Ralph Waldo Emerson

"God didn't make a mistake when He made you. You need to see yourself as God sees you."—Joel Osteen

"I distrust those people who know so well what God wants them to do, because I notice it always coincides with their own desires."—Susan B. Anthony

"The best remedy for those who are afraid, lonely or unhappy is to go outside, somewhere where they can be quiet, alone with the heavens, nature and God. Because only then does one feel that all is as it should be."—Anne Frank

"Call on God but row away from the rocks."—Indian Proverb

"Either we are alone in the universe, or we are not. Either thought is frightening."—Arthur C. Clarke

"God will prepare everything for our perfect happiness in heaven, and if it takes my dog being there, I believe he'll be there."—Billy Graham

"Small amounts of philosophy lead to atheism, but larger amounts bring us back to God."—Francis Bacon

"Your talent is God's gift to you. What you do with it is your gift back to God."—Leo Buscaglia

"Our passionate preoccupation with the sky, the stars, and a God somewhere in outer space is a homing impulse. We are drawn back to where we came from."—Eric Hoffer

"Live, so you do not have to look back and say: God, how I have wasted my life."—Elizabeth Kubler-Ross


"Anyone who does not regularly gaze up and see the wonder and glory of a dark night filled with countless stars loses a sense of their fundamental connectedness to the universe."—Brian Greene


"God is inaccessible light, surpassing every light that can be seen by us either through sense or through intellect."—Thomas Aquinas


"The clear light of science, we are often told, has abolished mystery, leaving only logic and reason. This is quite untrue. Science has removed the obscuring veil of mystery from many phenomena, much to the benefit of the human race: but it confronts us with a basic and universal mystery—the mystery of existence . . . But we must learn to accept it, and our existence as the one basic mystery."—Julian Huxley


"The American philosopher William James often said that religious experience emanates from 'feeling at home in the universe.' With bodies composed of particles derived from the birth of stellar bodies and containing organs shaped by the workings of planets, eroding rocks, and the actions of the seas, it is hard not to see home everywhere."—Neil Shuban, from his book "The Universe Within: The Deep History Of The Human Body".

"As to the cause of the universe, in context of expansion, that is left for the reader to insert, but our picture is incomplete without God."—Edward Arthur Milne, British cosmologist

"If we do discover a complete (unified) theory (of the universe) it should, in time, be understandable, in broad principle, by everyone, not just a few scientists. Then we shall all . . . be able to take part in the discussion of why it is that we and the universe exist. If we find the answer to that, it would be the ultimate triumph of human reason, for then we should know the mind of God."—Stephen Hawking, author of "A Brief History Of Time"

"The higher we climb the more comprehensive the view. Each new vantage point equals a better understanding of the interconnection of things. What is more, gradual accumulation of understanding is punctuated by sudden and startling enlargements of the horizon, as when we reach the brow of a hill and we see things never conceived of in the ascent."—Julian Barbour, commenting on the continual advancement of science.

"From the smallest necessity to the highest religious abstraction, from the wheel to the skyscraper, everything we are and everything we have comes from one attribute of man—the function of his reasoning mind."—Ayn Rand

"A great ball of fire about a mile in diameter, changing colors as it kept shooting upward, from deep purple to orange, expanding, growing bigger, rising as it was expanding, an elemental force freed from its bonds after being chained for billions of years."—William L Laurence, journalist, reporting on the first atomic explosion in the New Mexico desert, July 16, 1945.

"If you want my final opinion on the mystery of life and all that, I can give it to you in a nutshell. The universe is like a safe to which there is a combination. But the combination is locked up in the safe."—Peter De Vries

***Somehow***

I've tried for many an hour and a minute
To imagine this world without me in it.
I cannot think of a newborn day
Without me here—somehow—some way.
I can't imagine the autumn's flare
Without me here—alive—aware.
I cannot think of a dawn in spring
Without my heart's awakening.
These treasured years will come and go
With swifter pace, but this I know,
I have no fear—I have no dread
Of that marked day that lies ahead.
My flesh will turn to ash and clay
But I'll be here—somehow—some way.

--Don Blanding

"The creation of a thousand forests is in one acorn."—Ralph Waldo Emerson

"The tragedy of modern man is not that he knows less and less about the meaning of his own life, but that it bothers him less and less."—Václav Havel

"I am become Death, Shatterer of Worlds."—Robert J. Oppenheimer, leading scientist of the Manhattan project, which developed the atomic bomb, remembering the line from the Indian scripture Bhagavad-Gita, after witnessing the world's first nuclear explosion in the early morning darkness of July 16, 1945 in the desert of New Mexico.

"I still find each day too short for all the thoughts I want to think, all the walks I want to take, all the books I want to read, and all the friends I want to see. The longer I live the more my mind dwells open the beauty and the wonder of the world."—John Burroughs

"Death is not extinguishing the light; it is putting out the lamp because dawn has come."—Rabindranath Tagore

"God is the universal substance in existing things. He comprises all things. He is the fountain of all being. In Him exists everything that is."—Seneca

"Some cancer patients recover; some don't. But the ordeal of facing your mortality and feeling your frailty sharpens your perspective about life. You appreciate little things more ferociously. You grasp the mystical power of love. You feel the gravitational pull of faith. And you realize you have received a unique gift—a field of vision others don't have about the power of hope and limits of fear; a firm set of convictions about what really matters and what does not."—Tony Snow, former White House spokesman and news commentator, describing his view of life during his losing battle with cancer.

"The longer I live, the more beautiful life becomes."—Frank Lloyd Wright

"I went to the woods because I wished to live deliberately, to front only the essential facts of life, and see if I could not learn what it had to teach, and not, when I came to die, discover that I had not lived."—Henry Thoreau

"I am the captain of my soul. I am the master of my fate."—William Henley, from his poem "Invictus".

"If I had my life to live over, instead of wishing away nine months of pregnancy, I'd have cherished every moment and realized that the wonderment growing inside of me was the only chance in life to assist God in a miracle…I would have sat on the lawn and not worried about grass stains…When my kids kissed me impetuously I would never have said, 'Later. Now go get washed up for dinner."—Erma Bombeck

"Perhaps it would be a good idea, fantastic as it sounds, to muffle every telephone, stop every motor and halt all activity for one hour someday just to give people a chance to ponder for a few minutes on what it's all about, why they are living and what they really want."—James Truslow Adams

"Figuring out who you are is the whole point of human experience."—Anna Quindlen

"Mountains are the beginning and the end of all natural scenery."—John Ruskin

"If you see no reason to give thanks, the fault lies in yourself."—Native American proverb

We know the truth, not only by reason, but also by the heart."—Blaise Pascal

"Life is good whether it is long or short. I feel like going into the wild is a calling all feel, some answer and some die for."—Final texted words of Bryce Gillies as he lay dying, found on his Blackberry. Gillies, an experienced hiker, celebrating his 20th birthday, had gone solo hiking on a back country trail in the Grand Canyon during July 2009, took a wrong turn and later died from dehydration and sunstroke, after he found himself on a precipice 80 feet high, a half mile from the Colorado River, without the strength to return the tortuous way he had come. Gillies was one of 12 people who died in the Grand Canyon in 2009. (From a story in The Arizona Republic by Richard Ruelas).


"I did not wish to take a cabin passage, but rather to go before the mast and on the deck of the world, for there I could best see the moonlight amid the mountains."—Henry David Thoreau


"No man is an island, entire of itself; . . . any man's death diminishes me, because I am involved in mankind, and therefore never send to know for whom the bell tolls; it tolls for thee."—John Donne


"I don't want to get to the end of my life and find that I lived just the length of it. I want to have lived the width of it as well."—Diane Ackerman

"Strange, isn't it? Each man's life touches so many other lives, and when he isn't around he leaves an awful hole to fill, doesn't he?"—Clarence, George Bailey's guardian angel, from the movie "It's A Wonderful Life", after George had been shown how events would have transpired if George had never been born.

"Man is equally incapable of seeing the nothingness from which he emerges and the infinity in which he is engulfed."—Blaise Pascal

"There's something wonderfully inspiring about rivers—about the determination with which they set their sights on the sea and weave around the infinite obstacles in their path while never losing their sweet trilling song."—Kevin Grange, from his book "Beneath Blossom Rain".

"You've never lived until you've almost died. For those who fight for it, life has a meaning the protected will never know."—Leigh Wade

"To touch a child's face, a dog's smooth coat, a petaled flower, the rough surface of a rock, is to set up new orders of brain motion. To touch is to communicate."—James W. Angell

"A man has made at least a start in discovering the meaning of human life when he plants shade trees under which he knows he will never sit."—D. Elton Trueblood

"One of the greatest pleasures a parent can experience is to gaze upon the children when they're fast asleep."—Unknown

"Here is a test to find whether your mission on Earth is finished: If you're alive, it isn't."—Richard D. Bach

"Remember that what you possess in this world will be found on the day of your death to belong to somebody else. But what you are will be yours forever."—Henry Van Dyke

"Learn as if you were going to live forever. Live as if you were going to die tomorrow."—John Wooden

"A hundred years from now it will not matter what my bank account was, the sort of house I lived in, or the kind of car I drove. But the world may be different because I was important in the life of a boy."—Forest E. Witcraft, U. S. scouting administrator.

"No humane being, past the thoughtless age of boyhood, will wantonly murder any creature which holds its life by the same tenure that he does."—Henry David Thoreau

"As we survey all the evidence, the thought insistently arises that some supernatural agency - or, rather, Agency - must be involved. Is it possible that suddenly, without intending to, we have stumbled upon scientific proof of the existence of a Supreme Being? Was it God who stepped in and so providentially crafted the cosmos for our benefit?"—George Greenstein, astronomer

"Were there no God, we would be in this glorious world with grateful hearts with no one to thank."—Unknown

"Believe that life is worth living, and your belief will help create the fact."—William James

"Nothing revives the past so completely as a smell that was once associated with it."—Vladimir Nabokov

"The more we get to know about our universe the more the hypothesis that there is a Creator . . . gains in credibility as the best explanation as why we are here."—Dr. John Lennox, Oxford professor

"Someone once said that if you sat a million monkeys at a million typewriters for a million years, one of them would eventually type out all of Hamlet by chance. But when we find the text of Hamlet, we don't wonder whether it came from chance and monkeys. Why then does the atheist use that incredibly improbable explanation for the universe? Clearly, because it is his only chance of remaining an atheist. At this point we need a psychological explanation of the atheist rather than a logical explanation of the universe."—Peter Kreeft


"There is no silence like that of the mountains."—Guy Butler


"There can never be any real opposition between religion and science; for the one is the complement of the other. Every serious and reflective person realizes, I think, that the religious element in his nature must be recognized and cultivated if all the powers of the human soul are to act together in perfect balance and harmony. And indeed it was not by accident that the greatest thinkers of all ages were deeply religious souls."—Max Planck, Nobel Prize winning physicist


"Some people complain because God puts thorns on roses, while others praise God for putting roses among thorns."—Unknown

"A scientific discovery is also a religious discovery. There is no conflict between science and religion. Our knowledge of God is made larger with every discovery we make about the world."—Joseph H. Taylor, Jr., who received the 1993 Nobel Prize in Physics


"It may seem bizarre, but in my opinion science offers a surer path to God than religion."—Paul Davies, Physicist


"God exists whether or not men may choose to believe in Him. The reason why many people do not believe in God is not so much that it is intellectually impossible to believe in God, but because belief in God forces that thoughtful person to face the fact that he is accountable to such a God."—Robert Laidlaw


"An honest man, armed with all the knowledge available to us now, could only state, in some sense, the origin of life appears at the moment to be almost a miracle, so many are the conditions which would have had to been satisfied to get it going."—Francis Crick, Nobel prize winner for the co-discovery of DNA


"The universe in some sense must have known we were coming."—Freeman Dyson, physicist, commenting on the fact of the fine tuning of the universe that supports man's existence.

"This new realm of molecular genetics is where we see the most compelling evidence of design on the Earth."—Dean Kenyon, biology professor

"(Without an Intelligent Designer) the probabilities of forming a rather short functioning protein at random would be one chance in $10^{125}$—that's a ten with 125 zeros behind it! And that would only be one protein molecule—a minimally complex cell would need between three hundred and five hundred protein molecules. Plus all of this would have to be accomplished in a mere 100 million years, which is the approximate window of time between the Earth cooling and the first microfossils we've found. To suggest chance against these odds is really to invoke a naturalistic miracle. It's a confession of ignorance. And since the 1960s scientists have been very reluctant to say that chance played any significant role in the origin of DNA or proteins."—Stephen Meyer, PhD, philosopher and scientist

# REFERENCE SOURCES

## BOOKS

Behe, Michael, "Darwin's Black Box: The Biochemical Challenge to Evolution", Free Press, 2006 (Discussion of Irreducible Complexity)

Berman, Bob, "The Sun's Heartbeat", Black Bay Books, 2012

Bodanis, David, "$E=mc^2$: A Biography of the World's Most Famous Equation", Berkley Publishing Group, 2001

Bodanis, David E., "The Secret House—The Extraordinary Science of an Ordinary Day", Berkley Trade, 2003

Canfield, Donald E., "Oxygen: A Four Billion Year History", Princeton University Press, 2014

Comins, Neil F., "What If the Moon Didn't Exist? Voyages to Earths That Might Have Been and Discovering the Universe", iUniverse, 2012

Dawkins, Richard, "The Blind Watchmaker", W. W. Norton and Company, 1996

Dembski, William, "The Design Revolution: Answering the Toughest Questions About Intelligent Design", Intervarsity Press, 2004

Diamandis, Peter, Kotler, Steven, "Abundance: The Future Is Brighter Than You Think", Free Press, 2012

Eban Alexander, M. D., "Proof of Heaven: A Neurosurgeon's Journey into the Afterlife", Simon and Schuster, 2012

Fishman, Charles, "The Big Thirst: The Secret Life and Turbulent Future of Water", Free Press, 2012

Flew, Anthony, "There Is A God": How The World's Most Notorious Atheist Changed His Mind", HarperOne, 2008

Greenstein, George, "The Symbiotic Universe: Life and Mind in the Cosmos", William Morrow & Co., 1988

Gribbin, John, "Alone In The Universe: Why Our Planet Is Unique", Wiley, 2011

Gonzalez, Guillermo, Richards, Jay W., "The Privileged Planet: How Our Place in the Cosmos Is Designed for Discovery", Gateway Editions, 2020

Hawking, Stephen W., "The Theory of Everything", Jalco Publishing House, 2007

Hitchens, Christopher, "God Is Not Great: How Religion Poisons Everything", Twelve, 2009

Holt, Jim, "Why Does The World Exist? An Existential Detective Story", Liveright, 2013

Horn, Trent, "Answering Atheism", Catholic Answers Press, 2013

Mauceri, John, "For The Love Of Music: A Conductor's Guide To The Art of Listening", Knopf, 2019

Meyer, Stephen C., "Darwin's Doubt: The Explosive Origin of Animal Life and the Case for Intelligent Design", HarperOne, 2013

O'Leary, Densyne, "By Design Or By Chance? The Growing Controversy on the Origins of Life", Augsburg Books, 1984

Overman, Dean L., "A Case For The Existence Of God", Rowan and Littlefield, 2009

Pinker, Steven, "The Language Instinct: How The Mind Creates Language", York Press, 2001

Ridley, Matt, "Genome: The Autobiography Of A Species In 23 Chapters", Harper Perennial, 2006

Ross, Hugh , "The Creator and the Cosmos: How the Latest Scientific Discoveries of the Century Reveal God", NavPress Publishing Group, 2001

Russell, Bertrand, "Why I Am Not A Christian, and Other Essays on Religion and Related Subjects", Touchstone, 1967

Scharf, Caleb, "The Copernicus Complex: Our Cosmic Significance in a Universe of Planets and Probabilities", Scientific American, 2014

Shuban, Neil, "The Universe Within: The Deep History of the Human Body", Vintage, 2013

Strobel, Lee, "The Case For A Creator", Zondervan, 2004

Thaxton, Charles B, Bradley, Walter L., Olsen, Roger L., "The Mystery of Life's Origin", Dallas TX, Lewis and Stanley, 1984

Ward, D., Brownlee, David, "Rare Earth: Why Complex Life Is Uncommon In The Universe", Copernicus, 2003

Wells, Jonathan, "The Politically Incorrect Guide to Darwinism and Intelligent Design", Regency Publishing, Inc., 2006 (See pp 93-94 for a hilarious account of typing monkeys)


Websites:

http://www.arn.org

www.discovery.org/id/

http://privlegedspecies. com

www.thenakedscientists.com   (Source for the estimate of the number of atoms in one grain of sand)

www.trueorigin.org/language01.asp

https://www.khanacademy.org/science/cosmology-and-astronomy/universe-scale-topic

www.sirc.org/publik/smell.pdf (The Smell Report by Kate Fox)

www.ideacenter.org (Intelligent Design and Evolution Awareness Center)

www.emc2-explained.info/index.htm

http://en.wikipedia.org/wiki/Blood

http://zebu.uoregon.edu/~soper/Sun/fusionsteps.html (A step by step narrative explaining the sun's fusion process)

http://www.godandscience.org/apologetics/quotes.html

www.wikipedia (Source for many numerical amounts used in the text)


Other Media:

The Intelligent Design Collection (3 DVDs), Illustra Media, (www.illustramedia.com), 2002

Metaxas, Eric, "Science Increasingly Makes the Case for God", Commentary in The Wall Street Journal, December 26, 2014

www.ingramcontent.com/pod-product-compliance
Lightning Source LLC
Chambersburg PA
CBHW020014050426
42450CB00005B/469